A Systems-Thinking Model to Grow Local Living Economies at For the Children Inc.

ENLARGED PRINT EDITION

By Activist Author

Diane Freaney

Local Living Economies Project originally submitted: June 3, 2013

MBA Sustainable Systems, Bainbridge Graduate Institute (BGI), Stuart Cowan PhD, UC Berkley, Faculty Advisor

ISBN: 978-1-957365-08-4 (paperback)

Cover Photo by Felecia Hatcher

Author Photo by Carson's Tavern

Published by Emerson House Press

www.dianefreaney.com

Prologue

Back to my friends at the Wall Street Journal to help me assess the situation

> "Who Assassinated Haiti's President? The Mystery Gets Murkier." Drew Hinshaw, Ryan Dube, Kejal Vyas, and Juan Forero. Wall Street Journal. August 4, 2021. (Nearly a month after President Jovenel Moïse was killed, the circumstances are just as hard to parse, with more new questions than answers.)

My own naiveté is showing again. I believed that everything was going well because I wanted it to go well. I finally resigned as CFO of **For the Children Inc.** because I was being asked to do things that were not quite legal. It seems Miss Reggie wanted to live in Haiti half of the year and manage **For The Children** remotely.

She wanted me to go to Haiti to help secure the family business making rum. Why couldn't Miss Reggie do that? Her answer - She was a girl and her grandfather would never accept a girl as a business partner. I am a girl – I guess she didn't notice -and I had ZERO interest in going to Haiti. Finally she stopped nagging me.

Next she saw Miss Cynthia running very fast through a neighborhood of known drug dealers. Would I talk to Miss Cynthia? Why not Miss Reggie, Miss Cynthia's boss? I guess Miss Reggie was afraid of the drug dealers. I am also afraid of drug dealers.

The building, which had to be built by 2015 to maintain the lease for $1.00 a year is still not built – no wonderful commercial kitchen, with all the activities for children and youth and families.

Opioids invaded the neighborhood, taking the lives of students and teachers, destroying families. Then came COVID and forced lockdowns which put the poorest children into unhealthy home situations 24/7, with no respite during in-person school.

~ ~ ~

I could go on and on but I must stop. It seems that home-schooling hit Malcolm Gladwell's tipping point (10%) and went sailing past it to 15% to 20%. The old in-person school is done. Public school education as we know it is in its death throes, as are so many Big Institutions.

Table of Contents

1.0 Executive Summary

The year 2013 has been part of a long journey. I met Miss Reggie in 1999, and was captivated by her smile and passion to help "her people," particularly "her children". I have been the CFO (part-time) with **For The Children, Inc. (FTC)** since 1999. I oftenthought that I "should" be able to help more, but my skills were "command and control" and my solutions were "linear thinking." My solutions didn't work.

Miss Reggie is a "systems thinker." Miss Reggie has always known that you can only help a child, when you involve parents, the extended family, the neighborhood, theschools, the entire community. How do you stay true to your visionand continue to exist in a "linear thinking" world?

"Her people" are low income, working poor, undocumented, migrant workers, ESL(English as a Second Language), marginalized, underserved. These are words that the government agencies and foundations use to describe people that they fund.

This process creates dependency and a sense of hopelessness. Right now Miss Reggie's community needs public funding.

The goal of our project is to help the community move toward economic independence, not to produce a specific outcome.

Recently, I am seeing more hopefulness, more community members recognizing their own capabilities and assets. This project is ongoing, not ending. I look forward to continuing to workwith Miss Reggie, FTC staff, parents, children, and the entire FTC community.

I appreciate my classmates Michael B. Maine and Murthy Srinivas for asking to become involved with **For The Children**. Their unique expertise and perspective have been invaluable. Michael and Murthy are each working on separate projectsand injecting possibility into the FTC community.

I appreciate everyone in the FTC community for welcoming Michael, Murthy and me into your world.

2.0 Organizational Summary

For The Children Inc. (FTC), "A Place Where Children Come First," is the focal pointfor the Haitian community in Lake Worth, Florida (Palm Beach County). Reginale Durandisse (Miss Reggie) founded **FTC** in 1999 as a childcare agency, initially serving 25 students primarily from the Osborne neighborhood. Reggie's original vision is still thevision for the agency:

> Our vision is to enable children to reach their full potential by offering a safe,caring, educational environment in which children can grow emotionally, physically, socially, and intellectually.
>
> We believe that children are gifts that lead us into the future. They allow us to share their world. They enrich our lives as we attempt to enrich theirs. They trust us to nurture their bodies and their souls, to keep them safe from harm and support them through life's challenges. **At For The Children Inc.**, we are committed to helping our children reach their full potential. **FTC** serves children age three to high school and provides family support services for the entire community.

Families need a healthy and vibrant local economy, safe homes, supportive friends, neighbors, churches and other institutions and good jobs in their community so theyare able to provide the experience they want for their children. FTC has come a long way since 1999 and there is still a long way to go. This is a plan for the next ten years.

3.0 Project Scope, Strategy and Story

To understand **FTC**'s place in the community and the challenges to achieving itsvision, it is important to understand place in the context from an historical perceptive of Palm Beach County.

Palm Beach County was created in 1909. It was named for its first settled community, the Town of Palm Beach ("the Island"). Among the County's first residents were former slaves who escaped plantations located in Alabama, Georgia, and South Carolina. Former slaves provided significant labor for the building of the county, its hotels, houses, and Flagler's railroad.

The **Town of Palm Beach ("the Island")** is separated from the neighboring cities ofWest Palm Beach and Lake Worth by the Intercoastal Waterway. Henry Morrison Flagler, a founder of Standard Oil, made "the Island" accessible via his Florida East Coast Railway.

Flagler established "the Island" as a resort, anchored by his two luxury resort hotels,the Royal Poinciana Hotel and The Breakers Hotel. The beneficiaries of the Gilded Age (1877 – 1893) bought Flagler's house lots. In 1902, Flagler built a Beaux-Arts mansion and helped establish the Palm Beach winter ("the season") by constantlyentertaining.

An area known as the Styx housed many of the servants, most of whom were black. The workers rented their small houses from the landowners. In the early 1900s, the landowners agreed to evict all Styx residents and raze the houses. Former Styx residents moved to West Palm Beach, a service town across the Intercoastal Waterway from "the Island."

The first settlers in the **City of Lake Worth** were Samuel and Fannie James, ex-slaves who claimed their land on the shores of the Lake Worth Lagoon under the Homestead Act in 1883. In the 1920s, farmers from other parts of the American south and mid-west relocated to Lake Worth to take advantage of the winter vegetable markets. After World War II with many modest pensioners from Quebec and Finland moved to the city.

During the past three decades, immigration from Central America and Mexico increased the Hispanic population. Included in the 1980s immigration were Guatemalan-Mayas, indigenous peoples who speak M'am, Q'anjob'al, or 22 other indigenous languages, and a large Haitian population, speaking Creole and French.

Belle Glade is the largest city within the 2,862,000-acre subtropical Everglades.

Belle Glade's motto is "Her Soil is Her Fortune." Belle Glade is often referred to as "Muck City" because of the rich black soil, which is perfect for growing sugar cane and vegetables. Today's crops are sugar cane, sweet corn, rice, bell peppers, lettuce, radishes, Chinese vegetables, specialty leaf, green beans and celery.

Tourism, agriculture, and construction are the three major multibillion-dollar industries in Palm Beach County. East County, near the water, is focused on tourism and West County is focused on agriculture. Belle Glade City Hall is 45 miles due west of the Town of Palm Beach Town Hall, yet many people living in "the Glades" have never seen the ocean and many residents of East County have never been to "the Glades." Most County residents have little or no understanding of the importance of agriculture to the economic sustainability of Palm Beach County.

At the West Palm Beach Greenmarket, only 8 of the 92 vendors are Farmer/Growers of vegetables. Other local farmers markets may not have any Farmers/Growers and feel fortunate if they are able to attract one or more Produce Resellers. This strikes me as odd because Palm Beach County is the "Winter Vegetable Capital of the United States."

FTC's main office and the original site for the childcare program is the historic Osborne School in the Osborne neighborhood. The Osborne School was the only school that colored children could attend in Palm Beach County during segregation. To give some perspective, Palm Beach County was among the last school districts inthe nation to integrate in 1971.

The **Osborne neighborhood** is the original black neighborhood in Lake Worth, just the other side of the tracks and in close proximity to the only black school in the neighborhood. **FTC** staff member, Miss Trudy, grew up in the neighborhood and is the keeper of the oral history. Miss Trudy still lives across the street from the main office and she has seen the neighborhood change. The neighborhood deteriorated as other housing options opened up to black residents. Recent immigrants, primarily Guatemalans, moved in as did drug dealers and prostitution rings.

The City of Lake Worth leased the Osborne School to Miss Reggie and her two staff members for $1-a-year in 1998. The building had been derelict for many years and they spent the first month shoveling out trash, cleaning and painting to make it habitable for the 25 children in the program.

The neighborhood did not embrace their new neighbors. The parents did not willingly send their children to childcare. In their countries, children stayed home when their parents worked, with the older children in charge. The parents had to beconvinced of the motives of FTC's staff, and the value of the program even though FTC did not charge parents.

4.0 Key Internal Stakeholders & Collaborative Partners

Key internal stakeholders

- Reginale Durandisse, Founder/CEO
- FTC staff, their children and extended families
- FTC parents, their children and extended families
- Osborne, Barton and adjacent neighborhoods in the City of Lake Worth
- Families with immigration issues residing in Palm Beach County
- Persons of Haitian decent residing in Palm Beach County
- Collaborative partners
- City of Lake Worth – city manager, mayor and commissioners
- Florida elected officials – Jeff Clemens and others
- Palm Beach County Commissioners – Grant Program Managers – Channel and Rene
- Central Palm Beach Chamber of Commerce – Nadine Burns
- PNC Bank – Dave Baker and Dave Vespo, Bank VPs
- Palm Beach County School Board – Curriculum Committee
- Lake Worth High School Principal
- Barton Elementary School Principal

- Southern Waste System, LLC and Sun Recycling – Patti Hamilton (neighbors)
- Prime Time Palm Beach County
- Compass - LGBT Center in Lake Worth – Julie Seaver, Ana Rodriguez, Tony Plakas, Jamie Todd Foreman
- Children's Services Council
- Family Central
- Churches – Grant Chapel A.M.E. Church, St. Andrews Episcopal Church, others
- Mother Earth Coffee Shop – Patti Lucia
- Gratitude Leadership Classes – ML-6, ML-7, ML-8, ML-9, ML-10, ML-11

5.0 Project Implementation & Design Strategies

The frame for this project is Asset-Based Community Development (ABCD). ABCDbuilds on the skills of local residents, the power of local associations, and the supportive functions of local institutions, and draws upon existing community strengths to build stronger, more sustainable communities.

ABCD draws on the analogy of a glass, which is half full or half empty depending on your perspective. A neighborhood that is portrayed as resourceful and gifted will bemore likely to draw upon the skills of local people to address issues and solve problems. Outside assistance and resources may still be required but the local community must set the agenda. People in these neighborhoods are positioned as engaged citizens, capable of shaping their futures. [1]

For Miss Reggie and For The Children, the glass is always half-full.

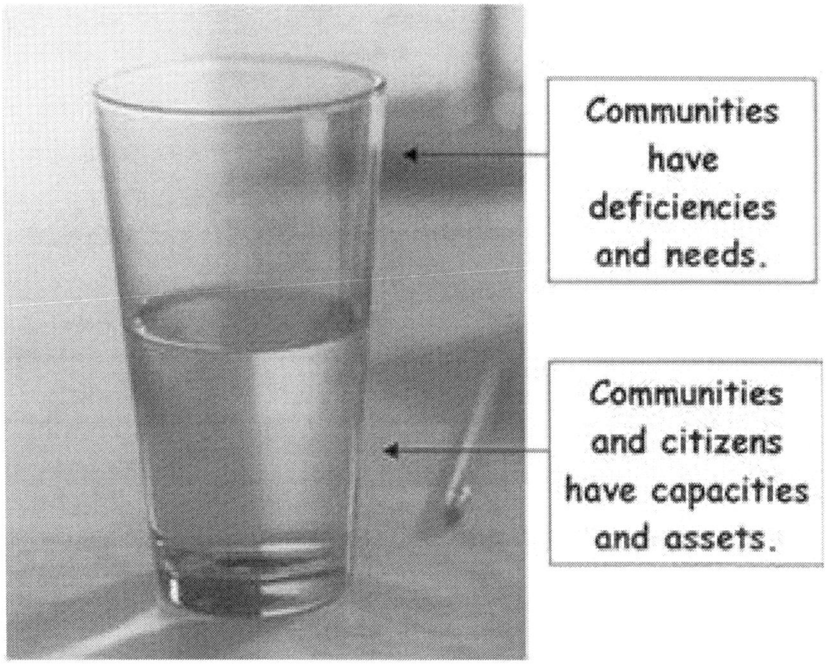

The "Street of Hope: The Rise and Fall of an Urban Neighborhood" is a model for change that provides inspiration for **FTC**. It is the story of the revitalization of the Dudley Street neighborhood. The Dudley Street Neighborhood Initiative (DNSI) wasformed in 1984 in the most impoverished neighborhood in Boston.

In 1984, Dudley Street was home to a culturally diverse group of residents. The residents were poor, often unemployed and on public assistance. The City of Bostonhad done a poor job of urban renewal and, while other areas of the city flourished, the Dudley Street area continued to go downhill.

The initial leadership came from the Riley Foundation. A Trustee made a long-term commitment, became embedded into the community, and pledged a significant portion of the Riley Foundation's grant budget for several years. From the beginning, the community was involved in the process. The leadership was recruited from the community and all decisions were debated and approved by the community. The Riley Trustee's connection to Boston's government agencies and the banks were extremely helpful in getting the community's concerns heard and then acted upon.

The book followed the Dudley Street project for ten years. The community made significant progress during that time and, in many ways, they were just beginning.[2] Today the DSNI web site celebrates many successes and still includes a long list of projects in the works.[3]

Of special significance, DSNI's Executive Director is John Barros, a lifelong resident of the Dudley neighborhood, selected after a national search in 2000. He has been involved since he was a teenager, serving as organizer, leader, and innovator in many ways. He was cofounder of DSNI's Nubian Roots Youth Committee, and helped start important DSNI youth programs such as college mentoring and job advocacy.

6.0 Project Plan & Timeline

My role is that of catalyst and connector, and works best when I am on site and embeddedin the community. My most important task is to be present, listen and let ideas/events/experiences emerge organically. When I am I Florida, I spent week days at FTC's main office. After hours, on the weekends, mornings and evenings, I attend community events in the FTC community, the City of Lake Worth, and otherareas in Palm Beach County.

Reggie and I also meet weekly on Skype and e-mail as needed on specific projects. The Internet culture is not as advanced in South Florida as it is in the Pacific Northwest and virtual meetings are less effective. Miss Reggie has said, "Ineed to see your eyes." Establishing trust is a difficult issue for FTC's community, primarily because many community members are undocumented.

The FTC community is part of a much larger system, as outlined above from an historical perspective. The system is constantly changing and adapting, as is any natural system. The FTC community has a broad understanding of what it wants to accomplish based on its vision. Within that understanding, priorities are constantly changing and adapting based on successes, failures, new alliances or new information.

The dilemma: How do we develop a sustainable business model that is independent of the need for public funds? The community has capabilities andassets. How do we tap into the community capabilities and assets?

At a second quarter intensive, Carol Sanford drew a line on the board. She labeled it, "Best Practices, standards, norms". Below the line, "Prevention, Problem Solving, and Punishment – 3 functions designed to keep us on the line". This pattern dominates in Palm Beach County, possibly a remnant of colonialism and the master/slave culture. This pattern also explains the hopelessness and despair that is so prevalent.

One of the goals of our work is to help community members tap into their own power and wisdom. The traditional approach has been to offer a "training" and teach people "how to do it right." The "training" approach builds dependency, and perpetuates the culture of hopelessness. How do we change the model?

7.0. Community Priorities and Resources

7A. Financial Matters

Miss Reggie identified "finance and budgeting" as a community problem. Individualcommunity members feel helpless to control their money. The mortgage crisis and the student loan problems heightened the sense of despair in the community.

Housing and Mortgages. Real estate agent and mortgage brokers targeted minoritypopulations, including FTC's community. Real estate agents and mortgage brokers were paid commissions based on deals that they closed, without regard for the ability of the buyer to pay the mortgage or maintain the property. For example, a Haitian woman with a salary of $18,000 a year was allowed to purchase a home for $500,000 with a 100% mortgage. She lost her home and moved in with relatives.

The Osborne neighborhood has many boarded up houses and empty lots. The LakeWorth CDC purchased lots in the neighborhood, built homes and helped first-time homebuyers through the mortgage process. This was during the run-up in housing prices and CDC homes sold for an average of $200,000.

The real estate bubble burst and home prices plummeted to $70,000 and less. Strapped homeowners, many lost their jobs during the downturn, stopped paying mortgages, and homes went into foreclosure. Absentee investors purchased the properties and rented them, usually to families from outside the Osborne neighborhood with values not consistent with the neighborhood culture.

FTC purchased 6 residential lots from the Lake Worth CDC for $15,000 on the street adjacent to Osborne School site. FTC also loaned Miss Trudy $10,000 to pay back taxes on her grandmother's home, the home that Miss Trudy grew up in and the oldest home in the original black neighborhood. Miss Trudy is paying back the loan through payroll deduction and the community is helping her with renovation, For example, someone showed up one day, cleared out overgrown bushes and painted the outside of the home.

Our hope and dream is to reclaim the community for its residents and take the community out of the Wall Street casino, perhaps through a community land trust. We will have to be patient. In Fall 2012, private equity firms moved into Palm Beach County and began "buying with the intent to rent."[4] Once again, prices are rising rapidly and homes are beyond the means of our families.

Miss Reggie advocates using a "rule of thumb" to pay no more than 3 times your income for a home. Our families are betting that Wall Street private equity firms will look for an "exit strategy" and start unloading homes, initiating another slide in real estate prices. Perhaps then homes will again become affordable for our families.

College and Student Loans Student loans are another issue weighing heavily on thecommunity. Miss Reggie has always encouraged children to excel in school and prepare for college. The community relied on college admissions officials and student financial aid officers to give guidance on selecting a college and financingtheir education. Many **FTC** children graduated from college and moved home saddled with student debt and unable to find a job.

We discovered that private colleges had lured many **FTC** students, with tuition andliving expenses funded entirely by "student financial aid," a euphemism for loans that must be paid back with interest.

Miss Reggie asked me to speak about the student loan problem on Haitian Radio. I don't speak Creole and, although I am passionate about solving the student loan problem, I do not have personal experience with student loans. I asked Carnee, FTC alum, college graduate, and mother of current FTC student Elijah and Meesha, FTC staff member and current Palm Beach State College student to join the conversation.

The radio show was scheduled for 1:00 PM; it was 1:15 PM and we were allsitting in the crowded community room at the Osborne Center. Suddenly Bob, Haitian radio show host, puts his cell phone in front of me and tells me to start.

After my introduction in English, Bob tells his cell phone back and translates into Creole. Carnee and Meesha do not speak Creole so the process was repeated many times during the hour. We had not rehearsed but the program flowed easily becausewe all spoke from the heart.

Our message was simple:

- Do not sign anything you don't understand; tell your children the same.
- Private colleges are high-priced and don't provide a good education.
- Palm Beach State College has many tuition free programs, work study jobs inthe community, and a good record of job placement after graduation

- **FTC** will provide support you in making these decisions.

The radio station received many calls asking to repeat this program and more programs like it. The parents want to help their children but they lack the knowledge and understanding. Miss Reggie and the **FTC** staff are quietly helpingparents make better decisions about college and student loans.

Personal Money Management and Budgeting

Miss Reggie has asked me "to teach finance" to her community for over ten years. Icouldn't envision how my middle class financial planning techniques could help a person who had trouble paying the rent and was food insecure.

Then I read **All Your Worth**,[5] a book that Elizabeth Warren wrote with her daughter. Warren says there are only three budget categories: Needs, Wants, and Savings.

Warren says that everyone deserves to have "Wants" even when he or she currentlydoes not have enough money to cover "Needs."

The book has a perspective of abundance (everyone deserves to have Wants), not deprivation. In a community where hopelessness is a huge problem, this is a dramatic shift in perspective. The book gives strategies to manage within the threecategories, with the objective of finding 30% for "Wants" and 20% for "Savings." "Savings" includes paying off credit card debt and student loans, a harsh reality for people living in poverty.

I bought some copies of **All Your Worth** and left them with Miss Reggie. One day I walked in to the main office and something felt different. I had offered to help Joe, Miss Trudy's son, solve his student loan problem and then not delivered on my promise. Joe came to tell me he had run his credit report. His credit report had one overdue bill for $200 and his student loans. He paid the $200 bill and made a plan topay his student loans by the end of the year. His fledging ice cream business was expanding and he wanted to buy a new car so he knew he had to improve his creditscore.

Miss Trudy and Miss Liz told me they had started bringing their lunch to work and were using the savings to pay overdue bills. Later Miss Trudy took me aside and toldme she had opened a savings account and began saving money. She leaned closer and whispered, "I have saved $2,000 so far."

Miss Reggie's children, Hazeller and Christina, and the other young people that they hang out with, had started an informal support group. Hazeller decided how much he could spend for "Wants" each paycheck and has the rest direct deposited into his savings account. The group often discusses potential purchases, and challenges each other if the purchase is a "Want" or a "Need." It seems like the conversation around money is changing.

Women and Finance Miss Trudy, Director of the Early Childhood Program, sharedher own story of taking control of her money in a workshop that she organized on May 24, 2013. (See the video by visiting **The Snarky Express YouTube Channel** and search for the Woman and Finance Part I video or go here: https://youtu.be/FLnaoczcWY8) Miss Trudy expected 20-25 people at the meeting; 32 adults and 20 children showed up. The discussion after Trudy's presentation was lively, the questions were good, and there are plans to continue meeting monthly to share stories and support each other.

When you watch the video, you will notice some men in the room. Hazeller and Mr. Vincent came to monitor a separate room for the children. The mothers decided they wanted to keep their children with them to share the conversation and the mealthat Miss Trudy and her friends prepared for the occasion.

7B. Food and Nutrition

Many members of FTC's community farmed or kept gardens in their countries before immigrating to Palm Beach County. Many settled in "the Glades" and followedthe crops as migrant farm workers, worked for local farmers in "the Glades", and then perhaps joined with friends or started their own little business as gardeners in gated communities.

Agriculture is the second largest industry in Palm Beach County and it is invisible tomany residents and most visitors. The mental model that many people have is of migrant farm workers abused by greedy, uncaring farmers. This model is obsolete today in most cases. This obsolete model robs both the farmers and the workers of satisfaction and pride in their work. The reality is that the farm owners and the farmworkers all work very hard to produce safe and nutritious food and deliver it fresh to our grocery stores and restaurants.

FTC's community is distant from the Glades and most community members do notwork in agriculture. Many need food assistance from SNAP (food stamps), WIC (Women, Infants, Children), or the School Lunch Program. Few families keep gardens.

FTC assists Community members in signing up for SNAP and WIC. FTC works with the Palm Beach County School District, the Palm Beach County Commissioners and the State of State to assure that the children have meals when school is in session.

The USDA mandates these programs and – on paper – they are designed to provide a nutritious diet.

7C. School Lunch

The Voluntary Pre-Kindergarten and the Early Childhood students participate in the USDA program administered by the State of Florida. The children receive breakfast, lunch, and a snack in the afternoon. FTC does not have a catering kitchen, only a food assembly room, so they must hire an approved caterer to provide the food. The Health department has to certify the food preparation room and every staff member who will be serving the food. FTC can select the caterer from a list of State of Florida authorized caterers.

The caterer delivers the food each morning in time for the State-designated lunchtime of 11:30 AM. The hot portion of the meal arrives in a heating unit. The cold portion may be in a can, plastic bags, and plastic gallons of milk or occasionally boxes. The food is served on paper or Styrofoam plates with plastic sporks (spoon/forks), and 4 ounces of milk in a plastic cup (required to be served with each meal or FTC does not get reimbursed).

The USDA sets the menu for each day in a five-week rotation. FTC has a choice of 5 possible set menus but that is the only choice. For two weeks, I observed the children eat, talked to the group leaders about the meal, and ate the same food as the children.

The staff knew the children's tastes and told me before each meal was served what the children would eat and what would end up in the trash. The same children would drink the milk each meal; the rest of the children threw it in the trash. The staff pointed out 10 of the 30 meals on the rotating 5-week menu that the children liked, 10 that most of the children would eat, and 10 that almost none of the children would touch.

The program also includes an educational component. The staff did an excellent job of imparting the knowledge to the children. The children could all tell me why the meal in front of them was good for them. One little boy told me how much he liked his lunch and why it was good for him. Then he picked up his uneaten lunch, walked over and threw it in the trash.

What did we learn from this program?

- Food is only nutritious if it is eaten
- Knowing what is good for you is not the solution
- Children are not connected to the source of their food
- Children are taught not to trust their bodies (drink your milk) or eat when hungry (11:30 AM lunch MUST be served)
- Environmental issues – paper, plastic, Styrofoam
- Wasted food – USDA doesn't count it
- USDA knows best – disconnects parents from providing nourishment for their children

Miss Reggie's vision is a full catering kitchen in a new community center. **FTC** can hire parents from the community to create meals and snacks for the children from fresh ingredients, including vegetables from **FTC**'s community garden. The **FTC** children will work along with their parents and other community members to plant, maintain and harvest the vegetables.

The community will meet monthly – or perhaps weekly - to share meals prepared in the catering kitchen by the children, other parents and other members of the community. The children will have weekly cooking lessons in the catering kitchen so they learn first hand where their food is coming from and how to prepare the food in ways that are not just nutritious but appetizing. The catering kitchen will have a dishwasher, recycling and composting facilities to teach children and adults ways to care for their environment. The teens can use the catering kitchen to prepare healthy snacks and non-alcoholic beverages for their weekend hangouts.

Current situation

- **FTC** will continue to participate in the School Lunch and other feeding programs for the school year and summer. **FTC** is grateful that the children are not hungry in class and trusts that the USDA and other agencies will improve the program.
- The Gratitude Leadership Groups have adopted the **FTC** community garden as their service projects. Each time they come they improve the garden and the harvest increases. The children help tend the garden and are encouraged to eat the vegetables from the garden.
- A group of **FTC** children, parents and staff took the **Farm City Tour** run by the Central Palm Beach Chamber of Commerce on November 30, 2012. In the morning, the tour showed them the sugar cane harvest (by burning the fields), and the trip to the sugar mill and the entire process of converting cane to the

sugar for our cereal. In the afternoon, they went to TKM Farms, the largest lettuce grower west of California and Roth Farms, growers of sweet corn, green beans, and radishes. Their final stop was a nursery and the owner gave each person a pepper plant to take back and plant to the school garden. (To see a video about this, visit **The Snarky Express YouTube Channell** and look for the Farm City Tour video or go here: https://youtu.be/JJqvKc-eO2Y)

- Plans are to send a group of children and parents every year on the Farm City Tour. For now, **FTC** plans to put the video and other pictures from the tour on **FTC**'s website.

Addendum: St. Andrews Episcopal Church, an **FTC** after school child care site, has acommercial kitchen, certified by the health department. Can we cook the food at St. Andrews and transport it to the Osborne School to serve to the children under our contract? We have determined that we can use vegetables grown in our garden sincethey are grown in raised beds. Our hope is to hire people from our community to cook culturally appropriate food - in our community that includes beans, rice, tortillas, plantains, fresh vegetables and fruit. Our plan is to continue to explore implementing this plan inthe coming year.

7D. Building Project

Building project - **FTC** must build a building on the 4.83 acres it leases from the city of Lake Worth for 99 years at $1.00 a year by July 2015 to keep its lease.

Murthy suggested a SEED classroom (Risa Wolfe's LLE project) as a demonstration unit. Murthy has introduced Risa and Reggie and arranged a tour of the SEED classroom after graduation.

8.0 Evaluation/Assessment

I have no metrics, I didn't use "tools" to accomplish an objective. I was present and listened and let things emerge organically. Perhaps this was my biggest learning, allowing things to emerge organically. I am very good at setting goals and objectives, at making a plan, checking off deliverables, etc. - a linear thinking model - the result is always what I planned but not always the best outcome. Allowing things to evolve organically - a systems thinking model - is much more challenging intellectually - I just wanted to DO something. Emotionally it felt so much easier. What did I have todo with the process? The outcome will be what the community wants - not what I dictate - and a resilient outcome. Community members are learning that they can control outcomes; they are not victims.

9.0 Appendix

Resources - Print

"Shifting focus: Alternative Pathways for Communities and Economies." Latrobe City and Monash University, Traralgon and Melbourne, Victoria. 2001

Medoff, Peter and Sklar, Holly. "Streets of Hope: The Rise and Fall of an Urban Neighborhood." Boston, MA: South End Press. 1984.
(Special thanks to Rachel Williams for recommending this book.)

Miller, Kimberly. Wall Street moves onto Main Street: Meet your new landlord. Palm Beach Post. April 13, 2013.

Sanford, Carol (2011-02-23). The Responsible Business: Reimagining Sustainability and Success . John Wiley and Sons.

Warren, Elizabeth; Tyagi, Amelia Warren (2005-03-08). All Your Worth. Simon & Schuster, Inc.

Wheatley, Margaret J.; Frieze, Deborah (2011-04-11). Walk Out Walk On: A Learning Journey into Communities Daring to Live the Future Now. Berrett-Koehler Publishers.

Wheatley, Margaret J. So Far from Home . Berrett-Koehler Publishers. 2012.

Resources – Web Sites

Dudley Street neighborhood Initiative (http://www.dsni.org/)

For The Children Inc. (http://forthechildrenfirst.org/)

Michael B Maine - photo documentary - photos and videos from Michael B. Maine's visits to **For The Children**. Most of these pictures were taken on March 3, 2013, when Leadership Gratitude ML-11 and other volunteers showed up to help plan the garden. The large video at the bottom of the page is particularly moving and will be used as a focal point on the **FTC** website. http://michaelbmaine.com/ftc

Two Loops Model – Berkana Institute's Theory of Change – video presentation by Deborah Frieze on The Theory of Change https://youtu.be/ZcyHKKc2LVg

Resources - Videos

Farm Stand Field Trip - The children in Miss Cynthia and Miss Ruby's groups at the St. Andrew's Center planned a field trip to the local farm stand. The day before the field trip, the children chose teams and sat with their teams to plan their menu.

Each team had to include at least one child from the older group -- old enough to usea knife. Each team could choose four vegetables, three fruits, and a lettuce. The teams talked about possible substitutions in case the farm stand didn't have their first choice.

On the day of the field trip, Miss Cynthia and Miss Ruby drove the school vans to thefarm stand. All the children looked at the produce and helped make selections. The team captain was in charge of the basket and when the team was finished, the captain took the basket to the cashier. The cashier gave each team a receipt listingall their purchases and the cost.

Back at St. Andrew's Center, the teams arranged their purchases on their team tableand examined the cost of each item. One team was frugal and their purchases cameto $7.84. Another team purchased a pineapple, which brought their bill to $12.99.

The children took the vegetables into the kitchen and washed them. Then the younger children went back to their tables and drew vegetables and fruits on paperfor placemats. The children old enough to use knives prepared the fruits and vegetables for their feast. The children set the table and sat down to enjoy their feasttogether.

Farm City Tour - November 30, 2012 (To see video, visit **The Snarky Expresss YouTube Channel** or go here: https://youtu.be/QAmWQUjASpU) – Thank you to the following people who made this tour possible:

- Rick Roth, tour guide and farmer – Roth Farms
- American Sugar Alliance (https://sugaralliance.org/about-us)
- Jessica Clasby, tour coordinator and Membership Services & Young Professionals Manager – Central Palm Beach Chamber of Commerce
- TKM Bengard Farms LLC (https://www.facebook.com/TKMFarms/)
- Sugar Growers Co-Operative of Florida (http://www.scgc.org/)
- Florida Crystals (https://www.floridacrystals.com/our-story)
- The children, parents and staff of **For The Children Inc.** and especially Christina Decime who took this video.
- For The Children Garden Project - On February 28, 2012, Nadine Burns brought her Leadership class to **For the Children** to build a community garden - twelve raisedbeds. The group came with tools, soil, and plants - everything needed to make the garden a success.

The teens watered, weeded, nurtured, and finally harvested the produce from theirgarden. They had some challenged - bugs that liked collard greens eat all the plants in one day. The teens began keeping a notebook so they could identify the bugs and find a way to discourage them. For the collard green eating bugs - vinegar and watersprayed on the leaves.

July and August are too hot for vegetables so the garden is finished until the fall.

For The Children Talent Show - Tango (to see video, visit the Michael Maine YouTube Channel or go to http://youtu.be/BVx4trAxoyM) - 2012 Multicultural

Festival and Talent Show hosted by **For The Children, Inc.** Held at Barton Elementary in South Lake Worth, Florida on August 18, 2012 in memory of Alyx Kellington, the talent showfeatured youth performing song, dance, poetry, and gymnastics.

I Am An Author (to see video, visit the Reginale Durandisse YouTube Channel or go to https://youtu.be/PAIjxvKWLIA) - The children learn how to create their own books from beginning to end. Ms. Sharon Koskoff did a great job by showing our children that they can write and illustrate their own stories.

Lake Worth Middle School Garden – The garden club at Lake Worth Middle School proudly shows off their garden to visitors from **For The Children** in Lake Worth.

Today club members weed the garden and then pick vegetables to take home fordinner. James Quillian, PE teacher and garden club advisor, proudly shows off thestudents' work. The gardening project is so popular that all four PE classes have requested time in the garden.

Woman and Finance Workshop - May 24, 2013 (to see video, visit **The Snarky Express YouTube Channel** or go here: https://youtu.be/FLnaoczcWY8) – special thanks to Miss Trudy forarranging this meeting.

Resources – Still Pictures

One of the 30 lunches on the USDA 5 week menu rotation

Breakfast is served. This day the children had grape juice instead ofmilk. The milk is in the cereal.

Boarding the bus for the Farm City Tour

TKM Bengard Farms, LLC lettuce fields.

TKM Bengard Farms LLC Harvesting Lettuce

TKM Bengard Farms, LLC – Back on the Bus –Tasting Fresh

Burning the Sugarcane

Footnotes

[1] "Shifting focus: Alternative Pathways for Communities and Economies." Latrobe City and Monash University, Traralgon and Melbourne, Victoria. 2001

[2] Medoff, Peter and Sklar, Holly. "Streets of Hope: The Rise and Fall of an Urban Neighborhood." Boston, MA: South End Press. 1984.

[3] http://www.dsni.org/

[4] "Wall Street moves onto Main Street: Meet your new landlord," Kimberly Miller, PalmBeach Post, April 13, 2013

[5] Warren, Elizabeth; Tyagi, Amelia Warren. All Your Worth. Simon &Schuster, Inc. 2005

Epilogue

Elizabeth Warren, Harvard Bankruptcy Lawyer, should really have know better. Restoring bankruptcy rights for student debt would have gone a very long way in solving the student debt crisis. But it seems like Warren was intent on using the student debt crisis to promote her own political agenda.

And her books, **All Your Worth: The Ultimate Lifetime Money Plan** and **The Two-Income Trap: Why Middle-Class Parents Are (Still) Going Broke**, written with her daughter Amelia Warren Tyagi, Wharton MBA and Investment Banker, beg to be updated. What worked in 2005 does not work in 2021, probably hasn't worked for over ten years because of the nightmare of student financial debt. It pains me to see how much more difficult life has become for low-income marginalized folks.

The building, which had to be built by 2015 to maintain the lease for $1.00 a year, is still not built – no wonderful commercial kitchen, with all the activities for children and youth and families. The City of Lake Worth spent beaucoup bucks becoming the City of Lake Worth Beach , which has only managed to make housing, always unaffordable, less affordable for low-income folks in the Osborne Community.

The Surfside Condo Tragedy in Miami-Dade and the pandemic changed so much in 2020-2021.

What will 2022 bring? Most folks are tired of same old, same old and long for a different lifestyle.

Stay tuned.....

About the Author

I have been seated at the table – rather, more accurately, been seated behind the white men seated at the table and told to hold my tongue – at the launch of some of the most radical new business models of the last century.

I have had a front row seat for every new finance and/or economic theory that came down the pike.

This is a dubious honor. I have watched these same new business models crash and burn, take jobs, destroy families, make ghost towns of cities, compromise our health and well being, and rob us of our happiness. With the economic meltdown of 2008, I watched as my retirement fund plummeted 42% percent from the top of the market in 2007 to the bottom in 2009.

My life has been magical. I have over 70 years life/work experience and an excellent educational background.

Syracuse University (1965) – In my accounting courses, I learned to fill in forms and play games with numbers. In anthropology and public speaking, I learned storytelling and gained an appreciation of other cultures.

Harvard Business School – Corporate Finance Executive Education (1982) – I learned about OPM (Other People's Money), the strategy that brought the global financial markets to their knees in 2008.

University of Pennsylvania – Organizational Dynamics (1999) – I learned that student's work is only valued when it follows a structured academic path. My master's thesis, A New Model for the Creative Use of College Endowments to Reduce College Tuition, would have prevented today's student loan crisis. Penn had no mechanism for "the administrators" to listen to students.

Bainbridge Graduate Institute (2013) – I learned the importance of social media to listening deeply and delivering my message.

At an early age, I learned to communicate by listening. At my current age, I feel driven to share the knowledge and understanding amassed during my lifetime. Now I am speaking out.

—Diane Freaney, The Cat Lady

Made in United States
Orlando, FL
05 November 2026

90230148R00031